Pravin Gautam

Bacterial Speck Disease of Tomato: An Insight into Host-Bacteria Interaction

GRIN Verlag

Bibliografische Information der Deutschen Nationalbibliothek:

Die Deutsche Bibliothek verzeichnet diese Publikation in der Deutschen National-
bibliografie; detaillierte bibliografische Daten sind im Internet über http://dnb.d-
nb.de/ abrufbar.

Dieses Werk sowie alle darin enthaltenen einzelnen Beiträge und Abbildungen
sind urheberrechtlich geschützt. Jede Verwertung, die nicht ausdrücklich vom
Urheberrechtsschutz zugelassen ist, bedarf der vorherigen Zustimmung des Verla-
ges. Das gilt insbesondere für Vervielfältigungen, Bearbeitungen, Übersetzungen,
Mikroverfilmungen, Auswertungen durch Datenbanken und für die Einspeicherung
und Verarbeitung in elektronische Systeme. Alle Rechte, auch die des auszugsweisen
Nachdrucks, der fotomechanischen Wiedergabe (einschließlich Mikrokopie) sowie
der Auswertung durch Datenbanken oder ähnliche Einrichtungen, vorbehalten.

Imprint:

Copyright © 2008 GRIN Verlag GmbH
Druck und Bindung: Books on Demand GmbH, Norderstedt Germany
ISBN: 978-3-656-01093-7

GRIN - Your knowledge has value

Der GRIN Verlag publiziert seit 1998 wissenschaftliche Arbeiten von Studenten, Hochschullehrern und anderen Akademikern als eBook und gedrucktes Buch. Die Verlagswebsite www.grin.com ist die ideale Plattform zur Veröffentlichung von Hausarbeiten, Abschlussarbeiten, wissenschaftlichen Aufsätzen, Dissertationen und Fachbüchern.

Visit us on the internet:

http://www.grin.com/

http://www.facebook.com/grincom

http://www.twitter.com/grin_com

Bacterial Speck Disease of Tomato: An Insight into Host-Bacteria Interaction

Pravin Gautam*
Department of Plant Pathology
University of Minnesota
St. Paul, MN 55108, USA

Summary

Pseudomonas syringae pv. *tomato* (*Pst*) is a common pathogen of tomato which causes bacterial speck disease. This disease serves as a useful model for studying the interactions of microbial pathogens and plants. Most gram-negative bacteria, including *Pst,* have type III secretion system (TTSS). Encoded by *hrp* gene clusters, the TTSS is used to deliver effector proteins into the host cytosol. The *hrp* genes also control the expression of the avirulence genes (*avr*). One Avr protein, AvrPto, functions as ligand to elicit a hypersensitive response (HR) in the tomato plant after recognition by the protein encoded by the host resistance gene, *Pto*. The AvrPto-*Pto* interaction is the most widely studied systems. It has been discovered that *Pto* is linked with *Fen*, the gene responsible for susceptibility to an organophosphate insecticide, fenthion. Functioning of *Pto* requires another gene called *Prf*, which lies embedded in *Pto*. Though the system is well characterized, several aspects are still not understood. With the availability of completed genome sequence of *Pst* and the full sequence of tomato expected in the future, we may anticipate that our understanding of the mechanisms of this host-pathogen interaction to be improved.

Biological aspect of interaction

Pseudomonas syringae pv. *tomato* (*Pst*) causes bacterial speck disease of tomato wherever tomatoes are grown (8). Primary sites of infection are stomata, the bases of leaf trichomes and wounds (2). Following infection, bacteria multiply in the leaf interior by forming microcolonies in close physical association with the cell wall of host mesophyll cells (15). Once the host intercellular spaces are filled with *Pst*, host cells are polymerized and degenerate (37). Disease symptoms may be evident on all aboveground plant parts, though immature tissues are the most susceptible. Symptoms on leaves are often indistinct. Leaf spots are dark and round, and often have a discrete halo. As the disease progresses, lesions may extend into the petiole and stems. On fruit, the disease initially appears as small black spots of 1/8 – 1/4 inch diameter with distinct margins. These small spots are superficial, do not rupture the epidermis and will not develop into soft rot. Lesions on fruit are sometimes surrounded by an area that is slow to ripen. When fruits are infected early, the spots may cause pit-like distortions because the host tissues within lesions grow slower than unaffected tissue. Mature fruits are resistant to *Pst* infection as a result of their high acidity (8,38,42). Formation of halo in speck symptoms is due to the toxin, coronatine (COR), produced by *Pst* (3). Serious disease outbreaks, though rare, are favored by high leaf wetness, cool temperature and cultural practices that allow bacteria to be disseminated between hosts (26). Though the economic impact of disease is minimal, it is important from the point of view of scientific study of host-pathogen interactions. When *Pst* infects susceptible tomato plants it causes typical disease symptoms. In

* Author current address:
Small Grains Pathology Lab, Department of
Plant Science, South Dakota State University
Brookings, SD 57007, USA

contrast, infection of a resistant plant is restricted by the localized death of the cells at the site of infection known as the hypersensitive response (HR). The HR is generally microscopic but when it occurs over larger areas, becomes macroscopic. In addition, it has been reported that *Pst* can cause HR in non-host plants, which has been described as type II non-host resistance and shares some mechanistic similarities to the HR defense mediated by resistance genes (13,17).

Evidence for Communication between Microbe and Plant or vise versa-

The first action required for successful infection by pathogen is entrance into the host plant. The stomatal entry is important for *Pst* and is aided by the toxin, COR (36). Under normal conditions plant stomata close, in response to signals perceived by guard cells in the presence of live bacteria, to limit bacterial entry. Stomata close upon detection of pathogen associated molecular patterns (PAMPs). PAMPs are mostly lipopolysaccharides (LPS) and flagellin (flg22) which are identified by an unknown immune receptor and flagellin receptor, FLS2, respectively. The pathway for stomatal closure involves triggering of the salicylic acid (SA) and abscisic acid (ABA) signaling pathways (15). However COR produced by *Pst* in the apoplast or on the leaf surfaces re-opens closed stomata thereby increasing the liklihood of *Pst* invasion. Malic acid and citric acid, with minor contributions from shikimic acid and quinic acid, released from the plant tissues induce COR production (16). The COR promotes stomatal reopening through the E3 ligase subunit COI1, a key component of jasmonic acid (JA) signaling pathway. How *Pst* recognizes open and closed stomata is not clear. It is speculated that *Pst* movement and detection of stomatal status is chemotaxic due to nutrients released from interior of leaves through open stomata. Thus COR is an important virulence factor, which promotes colonization of host tissue by suppression of PAMPs induced early defense response in tomato (15,20,21).

Pst and other plant pathogenic bacteria have evolved a variety of virulence factors to avoid or supress host defenses and colonize plants successfully. One such virulence factor is the hypersensitive response and pathogenicity (*hrp*)-gene-encoded type III secretion system, (TTSS; see detail in molecular biology and genetics section). The TTSS is used by bacteria to inject a large number of virulence effector proteins into the host cell. Once *Pst* is inside the plant's apoplast, it needs signals for the induction of the *hrp* gene cluster. Mostly *hrp* gene expression is under the control of signals from the plant, and physiological and environmental conditions. In a study, when different *hrp* genes were expressed under varying conditions *in vitro*, optimum expression of all *hrp* genes were under conditions that simulate the apoplastic condition (low pH and limited nutrient condition: ie hypo-osmotic environment). Similarly, *hrpL* and *hrpRS*, members of *hrp* gene family, were actively expressed only in host plant leaves (27). Thus, once *Pst* are in the apoplast, the osmotic pressure, nutrient and pH condition of apoplast and some other signals from the plant induce the expression of *hrp* genes.

Physiological Alterations in the Participants

Several genes are induced both in *Pst* and tomato plants after infection. Especially, after the production of different effector proteins through the TTSS, there is up- and down-regulation of several genes that alter the physiology of host cells and induce the defense mechanisms.

These changes include, cell wall callus deposition, lignification of cell wall, oxidative burst, increase in the cell protectants, up-regulation of anthocyanin biosynthesis, JA and alkaloid biosynthesis, senescence, upregulation of defense related genes and ultimately the HR. Similarly, photosynthesis related activity, cell osmotic pressure related genes, and some protein kinases are down-regulated (Fig.1) (25).

Infection of tomato plants by *Pst* also alters the nitrogen metabolism of the host plant. Similar to abiotic stresses, *Pst* infection leads to reduced nitrogen assimilation in the infected cells. This has been shown by reduced expression of the primary nitrogen assimilation genes, *GS2* and *Nia* (24). Similarly, nitric oxide (NO), an important component for the induction of defense response, has been shown to be produced in a *hrp* dependent manner in host cells after *Pst* infection. NO production was preceded by the generation of hydrogen peroxide (H_2O_2) (22). However, NO production leading to basal defense has also been shown to occur in a *hrp* independent manner (36). Accumulation of a cytosolic glutamine synthetase (GS1) isoform, very similar to the GS1 isoform induced in tobacco leaves during senescence, was also observed (24). This indicates that the physiological processes induced during *Pst* infection are similar to those involved in natural senescence. It is interesting fact that several pathogen-related (PR) defensive proteins are expressed also in natural senescence. Wright and Beattie (40) observed that there is change in osmotic pressure of host cells after *Pst* infection. *Pst* encountered lower potential in incompatible reaction than in compatible reaction in tobacco plants. Thus, a reduction in the water potential of cells may restrict the endophytic growth of *Pst*. All these changes, decrease in water potential, reduction in nitrogen assimilation, and physiology similar to natural senescence are possibly linked to the HR. There is a possibility that *Pst* changes the host cell physiology through production of the phytohormone auxin. It has been indicated by induction of a gene, *iaaL*, by *hrpL* which encodes indoleacetate-lysine ligase capable of producing an inactive form of auxin and indoleacitic acid (IAA) in *Pst* DC3000 (35).

Host cell physiology can change even before the translocation of effectors proteins into the host cytosol. As mentioned earlier, the TTSS serves as a virulence factor for *Pst*. However, TTSS alone does not appear to be sufficient for bacteria to cause disease. Thus, a variety of virulence factors, such as COR, are necessary for full virulence. These virulence factors affect host cell by altering physiology. For example due to COR, *Pst* is able to reopen the stomata that are closed in response to the PAMPs through JA signaling pathway.

When a gene resistance to *P. syringae* pv. *tomato* (*Pto*) was transferred to susceptible tomato plants through breeding, there was an unintended consequence. Resistant tomato plants were found to be susceptible to organophosphorus insecticide fenthion (29). Susceptibility to fenthion resulted into HR similar to the HR produced *Pto* mediated resistace in tomato plants (28). With genetic study it was discovered that *Pto* resistance and fenthion sensitivity are encoded by two different genes, *Pto* and *Fen* respectively. The *Fen* gene is a *Pto* homolog but not involved in AvrPto recognition (28,29). The fenthion susceptibility trait is widely used for breeding resistance to bacterial speck of tomato (25).

Molecular Biology and Molecular Genetics

Several gram negative bacteria including *Pst* have the TTSS, enabling them to cause disease in host plants (9). The TTSS is a simple hollow tube-like structure consisting of inner and outer membrane rings and a protruding filament; a hrp pilus. The TTSS pilus is similar to a flagellum. In *Pst*, the TTSS pilus is about 8 nm wide and several micrometers long. The TTSS is encoded by a cluster of more than 20 hypersensitive response and pathogenicity (*hrp*) genes, organized into several operons on either the chromosomes or plasmids of the bacterium (1). The function of hrp pilus is to serve as conduit to guide the translocation of the type III effector proteins directly into the host cells cytoplasm (5,12,10). Initially, the *hrp* genes encoding the TTSS were identified in a *Tn5* mutant *Pst* which was unable to cause HR and lost pathogenicity. Thus it was termed *hrp* (HR and pathogenicity) and is synonymous to the plant pathogenic bacteria TTSS systems. However, the *hrp* system was found to be homologous to other TTSS systems, thus the nomenclature was changed to *hrc* (hrp conserved) (4). In the case of *Pst*, the *hrp* cluster is composed of 27 genes and arranged in four major operons, located on the chromosome (10). There are at least three classes of *hrp* genes at the *hrp* locus. The first class is the core component which codes for actual components of the TTSS. The second class are those genes; *hrpK*, *hrpL*, and *hrpS* that code for regulatory proteins (13). The third class encodes secreted proteins including an extracellular protein (HrpA).

The core *hrp* cluster (Fig 2.) is flanked by *hrpK* and *hrpL* on one side, and *hrpS* and *hrpR* on the other side. These are in turn flanked by a conserved effector locus (CEL) and an exchangeable effector locus (EEL) (10). The CEL contains at least seven open reading frames (ORFs), which are conserved in *Pseudomonas*, *Erwinia* spp and some other bacteria. The EEL has ORFs that even are not conserved in other *P. syringae* pathovars (26). Induction of *hrp* genes is regulated by *HrpL* through direct binding of the HrpL protein in the hrp box motif found in *hrp* genes (41,23). However, the induction of *hrpL* requires *hrpS* and *hrpR*. The HrpR and HrpS proteins forms a heterodimer on the *hrpL* promoter to stimulate the transcription of *hrpL* by interacting with the RpoN-RNA polymerase holoenzyme. Lon protease, HrpV and HrpA modulate the *hrpL* expression via unidentified post-transcriptional or transcriptional mechanisms (13). A two component system, GacS/GacA, (GacS is the sensory histidine kinase; GacA is the cognate response regulator) and HrpA (major component of type III pilus) play roles in regulation of *hrpRS* (35). Among the core *hrp* genes, expression of *hrpA* encodes proteins for pilus formation. As the pilus is assembling, the third class of *hrp* proteins; HrpZ, HrpW, are passed through the pilus (5,9,12). How the TTSS pilus penetrate plant cell wall and plasma membrane is not clear, but it is speculated that the extracellular HrpZ and HrpW proteins are responsible for forming pores in the plant cell membrane lipid bilayers (22). Once the TTSS pilus connects with the cytosol of the host cell, other *hrp* secreted proteins, virulence factors, and Avr proteins are translocated (Fig 3).

In many cases disease resistance responses in plants are elicited by the pathogen in a gene-for-gene manner as proposed by H. H. Flor, where the protein product of an avirulence (*avr*) gene from the pathogen is recognized by the protein from corresponding resistance (*R*) gene in the host plant (39). The interaction of *Pst* and tomato plants follows the gene-for-gene model where the *Pto* gene product in tomato plants is responsible for activating the defense system after recognition of the *avrPto* gene product from *Pst* (29,32). However, these *avr* genes are dependent on the *hrp* system for their functioning (25). AvrPto of *Pst* suppresses the expression of a suite of genes in *Arabidopsis* putatively involved in a cell-wall based defense response that limits the growth of TTSS deficit strains. This finding is supported by the suppression of cell-wall callus deposition and enhancement of population growth of a *hrp* mutant of *Pst* in transgenic plants expressing AvrPto. This suggests that AvrPto might function as a virulence factor (39). The *avrPto* is present in avirulent race 0 and absent in virulent race 1 of *Pst*, while its homologous genes are found in other pathovar of *P. syringae* which can attack bean, tobacco, pea, radish and oat (29), which suggests that *avrPto* is also recognized by *Pto*-like genes in other plants. The expression of *avrPto* is regulated by the hrp box in its promoter and induced under the conditions that stimulate the apoplastic space of plant leaves (31). The *avrPto* encodes an 18.3 kDa hydrophilic protein which has virulence activity and increases *Pst* growth in plants lacking *Pto* (7,33). Later it was discovered that *Pst* encodes another Avr protein, AvrPtoB, encoded by *avrPtoB*, which also interacts with an identical spectrum of *Pto* variants and elicits Pto mediated HR in tomato. However, unlike AvrPto, AvrPtoB is of larger size (56 kDa) and lacks myristylation motifs. Though the AvrPto and AvrPtoB are different proteins, they have common *Pto*-mediated avirulence activity due to the similarity of amino acid sequences at nine sub-regions with similar spacing between them (14).

The locus of the resistance gene *Pto* was introgressed from wild species of tomato (*Lycopersicum pimpinellifoliumi*) into cultivated tomato (*L. esculentum*) during 1930s (25). It was the first cloned *R* gene that operates in a gene-for-gene manner and is present in resistant genotypes as one member of a family of six genes clustered within a 60 kb region. *Pto* is missing in susceptible genotypes, although the gene family is present (11). Interestingly, susceptible tomato plants might still recognize AvrPto, albeit weekly, because of the presence of *Pto* family member gene *LecPtoF* (7). The *Pto* gene encodes a 321 amino acids protein with kinase catalytic domain with serine-threonine specificity and lacks leucine-rich repeats and a transmembrane domain (19), indicating that the Pto protein is localized in cytoplasm.

Initially, it was thought that the *Pto* was the only component involved in *Pto*-mediated defense mechanisms. However, through mutational analysis of *Pto*, Salmeron et al. (32) discovered that another co-segregating gene, *Pto* resistance and fenthion sensitivity (*Prf*), is also required for *Pto*-mediated resistance. *Prf* lies embedded within *Pto* next to the *Fen* gene and encodes a protein of large size (209.7 kDa), consisting leucine-rich repeats (LRR) and a nucleotide binding-Apaf1 resistance protein, Ced4 (NB-ARC) domain and putative leucine-zipper like regions (30). It is also found in several plant species, as its homologs have been found in pepper, tobacco, bean, maize, oat and *Arabidopsis* (25).

The question remains, why does *Pst* send Avr proteins into plant cells in order to be recognized by the host defense system. It is possibile that the original function of the Avr proteins may have been to promote disease in susceptible plants that lack the corresponding disease resistance genes. The avirulence function of these proteins results from the ability of some plants to evolve recognition capability (disease resistance genes) as a defense mechanism (12). How the AvrPto and AvrPtoB proteins are recognized by *Pto* or *Prf* products is not fully understood. Earlier it was hypothesized, that *Pto* induce basal defense by interacting with other *Pto* related proteins such as Pti1, Pti4, Pti5, and Pti6. In order, to suppress basal defense, AvrPto evolved and it disrupts this function by binding to Pto,

5

indicating Pto is the virulence target of AvrPto. *Prf* was proposed to have evolved later to recognize the Pto-AvrPto complex and activate the defense response. This hypothesis is known as 'guard hypothesis'. The guard hypothesis is supported by the facts that Pto and AvrPto interact physically (34), that the activities of *Pto* and *Prf* are interdependent (30), and that *AvrPto* does have virulence activity even when expressed in *Pst* in tomato lines expressing *Pto* but lacking *Prf* (6). However, more recent evidence indicates that Pto-mediated recognition is not supported by the original hypothesis. The AvrPto-Pto system now seems more likely to be supported by the 'Affinity-enhancement' model (Fig. 4). According to this model, Pto and Prf both are present in the host as a weak complex, possibly at the plasma membrane, which induce basal defense without presence of AvrPto. Since, Pto evolved to recognize avirulence proteins, binding of AvrPto likely stabilizes the complex and increases the abundance of the complex thereby increasing basal and other defense activities (25).

In addition to the recognition of AvrPto, emerging evidence suggests that *Pto* might utilize several signaling pathways for downstream defense induction. Salicylic acid (SA) is an important component of defense signaling. SA-binding proteins (SABP3), have been detected and shown to have a positive role in HR development in tobacco expressing *Pto* and *avrPto*. However, the precise role of SAB3 is yet to be identified (25). Similarly, mitogen-activated protein kinase (MAPK) is also a possible component for downstream *Pto* signaling as a rapid increase in MAPK activity has been found following expression of *AvrPto* and *AvrPtoB* in *Pto* expressing host plant.

Though the *Pst* –tomato interaction is a model system to study the molecular basis of plant disease resistance and susceptibility, it is still a complex system and not fully understood. Full genome sequencing of *Pst* DC3000 was completed in 2003. This has further aided our understanding of the system. Tomato genome sequencing is in progress (27% completed) and after its completion, researchers will be able to further explore the molecular and genetic mechanism of *Pst*-tomato interaction and shed additional light on the basis of plant resistance and susceptibility.

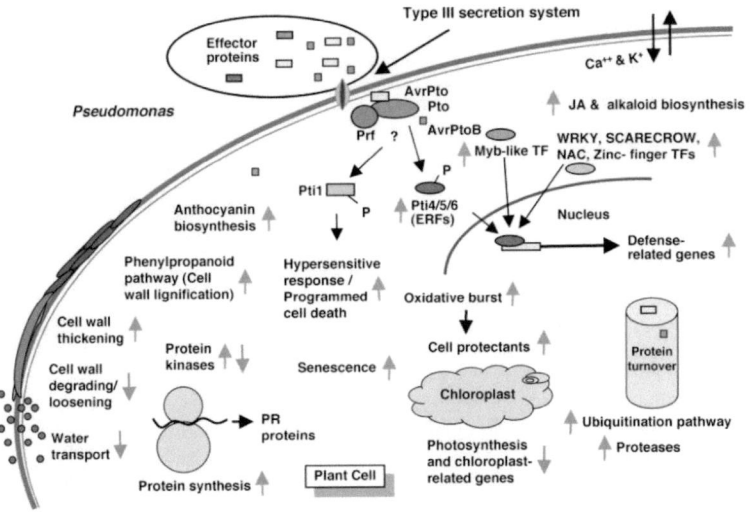

Fig. 1. Model of *Pto*- and *Prf*-mediated defense responses. Cellular processes suggested by gene expression profiling experiments to be involved in *Pto/Prf* mediated disease resistance. Green arrows indicate genes associated with the process that are induced and red arrows indicate genes that are suppressed. Black arrows indicate either possible steps in signal transduction pathways, flow of ions (for Ca+, K+), or host responses that might directly impact other responses (Pedley and Martin, 2003).

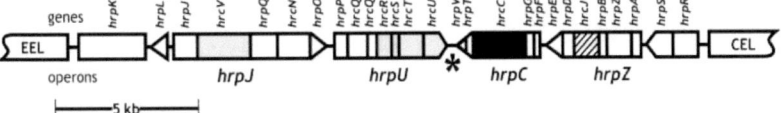

Fig. 2. Schematic of the core hrp cluster. Names above are genes and below are operon. The asterisk indicates the hypervariable region. CEL and EEL indicates conserved effectors locus and exchangeable effectors locus, respectively. Grey and black colored genes encodes protein associated with inner and outer membranes, respectively. Stripped is a gene associated with both the inner and outer membrane (Gropp & Guttman, 2004).

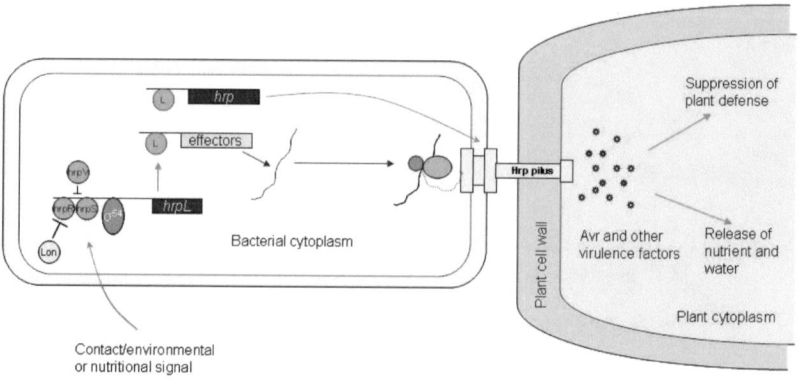

Fig. 3. Schematic of *Pst* TTSS. *HrpL /hrpS/hrpR* are stimulated by apoplastic conditions. The enhancer binding HrpR and HrpS binds to each other and activate σ^{54} dependent promoter of *hrpL*. Lon protease, HrpV and HrpA modulate the *hrpL* expression via unidentified posttranscriptional or transcriptional mechanisms. Expression of core *hrp* leads to assembly of hrp pilus across bacterial envelope, plant cell wall and plasma membrane. Avirulence protein and other virulence factors are translocated through Hrp pilus.(Adapted from Jin et al., 2003)

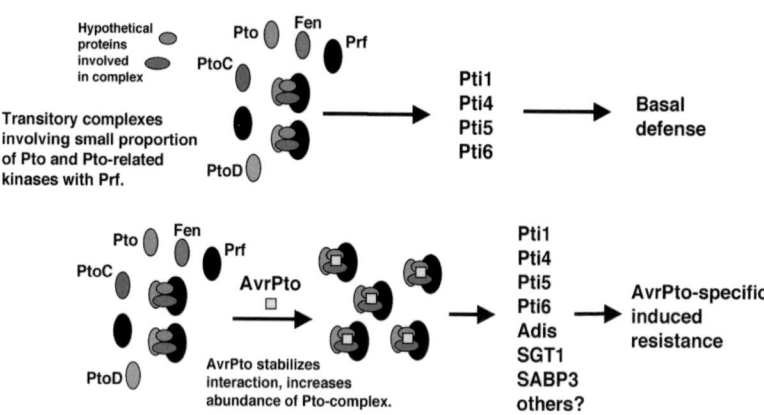

Fig. 4. The affinity-enhancement model. Prf is proposed to exist in transient complexes with Pto and other Pto-related kinases (e.g., Fen, PtoC, and PtoD). Additional proteins, shown in gray and brown, may also participate in this complex, which confers basal defense via Pti1, Pti4/5/6, and other proteins. Interaction of Pto and AvrPto is proposed to stabilize the Pto-Prf complex increasing its abundance. The stabilized complex acts via Pti1, Pti4/5/6, and additional proteins to activate AvrPto-specific induced resistance (Pedley and Martin, 2003).

References:

1. **Arnold DL, Pitman A, Jackson RW.** 2003. Pathogenicity and other genomic islands in plant pathogenic bacteria. *Molecular Plant Pathology* 4:407-420.

2. **Bashan Y, Sharon E, Okon Y, Henis Y.** 1981. Scanning electron and light microsocpy of infection and symptoms development in tomato leaves infected with *Pseudomonas syringae* pv. *tomato*. *Physiological Plant Pathology* 138:139-144.

3. **Bender CL, Stone HE, Sims JJ, Cooksey D.** 1987. Reduced pathogen fitness of *Pseudomonas syringae* pv. *tomato* Tn5 mutants defective in coronatine production. *Physiological and Molecular Plant Pathology* 30:273-283.

4. **Bogdanove AJ, Beer SV, Bonas U, Boucher CA, Collmer A, Coplin DL, Cornelis GR, Huang HC, Hutchenson SW, Panopoulos NJ, Gijsegem V.** 1996. Unified nomenclature of broadly conserved hrp genes of phytopathogenic bacteria. *Molecular Microbiology* 20:681-683.

5. **Brown IR, Mansfield JW, Taira S, Roine E, Romantschuk M.** 2001. Immunocytochrmical localization of HrpA and HrpZ supports a role for the Hrp pilus in the transfer of effector proteins from *Pseudomonas syringae* pv. *tomato* across the host plant cell wall. *Molecular Plant-Microbe Interactions* 14:394-404.

6. **Chang JH, Rathjen JP, Bernal AJ, Staskawicz BJ, Michelemore RW.** 2000. *avrPto* enhances growth and necrosis caused by *Pseudomonas syringae* pv. *tomato* in tomato lines lacking either *Pto* or *Prf*. *Molecular Plant-Microbe Interactions* 15:568-571.

7. **Chang JH, Tai YS, Bernal AJ, Lavelle DT, Staskawicz BJ, Michelemore RW.** 2002. Functional analysis of the *Pto* resistance gene family in tomato and the identification of a minor resistance determinant in susceptible haplotype. *Molecular Plant-Microbe Interactions* 15:281-291.

8. **Delahaut KA, Stevenson W.** 2004. Tomato and pepper disorders: Bacterial spot and speck. Cooperative Extension, University of Wisconsin, Publication *A2604*.

9. **Galan JE, Collmer A.** 1999. Type III secretion machines: Bacterial devices for protein delivery into host cells. *Science* 284:1322-1328.

10. **Gropp SJ, Guttman DS.** 2004. The PCR amplification and characterization of entire *Pseudomonas syringae* hrp/hrc clusters. *Molecular Plant Pathology* 5:137-140.

11. **Jia Y, Loh YT, Zhou J, Martin GB.** 1997. Alleles of Pto and Fen occur in bacterial speck-susceptible and fenthion-insensitive tomato cultivars and encode active protein kinases. *The Plant Cell* 9:61-73.

12. **Jin Q, He SY.** 2001. Role of Hrp pilus in type III protein secretion in *Pseudomonas suringae*. *Science* 294:2556-2558.

13. **Jin Q, Thilmony R, Zwiesler-Vollick J, He SY.** 2003. The III protein secretion in *Pseudomonas syringae*. *Microbes and Infection* 5:301-310.

14. **Kim YJ, Lin NC, Martin GB.** 2002. Two distinct *Pseudomonas* effector proteins interact with Pto kinase and activate plant immunity. *Cell* 109:589-598.

15. **Lefert PS, Robatzek S.** 2006. Plant pathogens trick guard cells into opening the gates. *Cell* 126:831-834.

16. **Li XZ, Starratt AN, Cupples DA.** 1998. Identification of tomato leaf factors that activate toxin gene expression in *Pseudomonas syringae* pv. *tomato* DC3000. *Phytopathology* 88:1094-1100.

17. **Lin NC, Martin GB.** 2007. Pto- and Prf- mediated AvrPto and AvrPtoB restricts the ability of diverse *Pseudomonas syringae* pathovars to infect tomato. *Molecular Plant-Microbe Interactions* 20:806-815.

18. **Lindgren PB.** 1997. The role of hrp genes during plant-bacterial interactions. *Annual Review of Phytopathology* 35:129-152.

19. **Martin GB, Brommonschenkel SH, Chunwongse J, Frary A, Ganal MW, Sipvey R, Earle ED, Tanksley SD.** 1993. Map-based cloning of protein kinase gene conferring disease resistance in tomato. *Science* 262:1432-1436.

20. **Melotto M, Underwood W, Koczan J, Nomura K, He SY.** 2006. Plant stomata function in innate immunity against bacterial invasion. *Cell* 126:969-980.

21. **Mittal SM, Davis KR.** 1995. Role of the phytotoxin coronatine in the infection of *Arabidopsis thaliana* by *Pseudomonas syringae* pv. *tomato*. *Molecular Plant-Microbe Interactions* 8:165-171.

22. **Mur, L. A. J., carver, T. L. W., Prats E.** 2006. NO way to live; the various roles of nitric oxide in plant-pathogen interactions. *Journal of Experimental Botany* 57:489-505.

23. **Nissan G, Manulis S, Weinthal DM, Sessa G, Barash I.** 2005. Analysis of promoters recognized by HrpL, an alternative sigma-factor protein from *Pantoea agglomerans* pv. *gypsophilae*. *Molecular Plant-Microbe Interactions* 18:634-643.

24. **Pageau K, Reisdorf-Cren M, Morot-Gaudry JF, Masclaux-Daubresse C.** 2006. The two senescence-related markers, GS1 (cytosolic glutamine synthetase) and GDH (glutamate dhydrogenase), involved in nitrogen mobilization, are differentially regulated during pathogen attack and by stress hormones and reactive oxygen species in *Nicotiana tabacum* L. leaves. *Journal of Experimental Botany* 57:547-557.

25. **Pedley KF, Martin GB.** 2003. Molecular basis of *Pto*-mediated resistance to bacterial speck disease in tomato. *Annual Review of Phytopathology* 41:215-243.

26. **Peterson GM.** 2000. *Pseudomonas syringae* pv. *tomato*: the right pathogen, of the right plant, at the tight time. *Molecular Plant Pathology* 1:263-275.

27. **Rahme LG, Mindrinos MN, Panopoulus NJ.** 1992. Plant and environmental sensory sigals control the expression of *hrp* genes in *Pseudomonas syringae* pv. *phaseolicola*. *Journal of Bacteriology* 174:3499-3507.

28. **Rommens CMT, Salmeron JM, Baulcombe DC, Staskawicz BJ.** 1995. Use of a gene expression system based on potato virus X to rapidly identify and characterize a tomato Pto homolog that controls fenthion sensitivity. *The Plant Cell* 7:249-257.

29. **Ronald PC, Salmeron JM, Carland FM, Staskawicz BJ.** 1992. The cloned avirulence gene *avrPto* induces disease resistance in tomato cultivars containing the *Pto* resistance gene. *Journal of Bacteriology* 174:1604-1611.

30. **Salmeron JM, Oldroyd GED, Rommens CMT, Scofield SR, Kim H, Lavelle DT, Dahlbeck D, Staskawicz BJ.** 1996. Tomato Prf Is a Member of the Leucine-Rich Repeat Class of Plant Disease Resistance Genes and Lies Embedded within the Pto Kinase Gene Cluster. *Cell* 86:123-133.

31. **Salmeron JM, Staskawicz BJ.** 1993. Molecular characterization and *hrp* dependence of the avirulence gene *avrPto* from *Pseudomonas syringae* pv. *tomato*. *Molecular and General Genetics* 239:6-16.

32. **Scofield SR, Tobias CM, Rathjen JP, Chang JH, Lavelle DT, Michelmore RW, Staskawicz BJ.** 1996. Molecular Basis of Gene-for-Gene Specificity in Bacterial Speck Disease of Tomato. *Science* 274:2063-2065.

33. **Shan L, He P, Zhou JM, Tang X.** 2000. A cluster of mutations disrupt the avirulence but not the virulence function of AvrPto. *Molecular Plant-Microbe Interactions* 13:592-598.

34. **Tang X, Frederick RD, Zhou J, Halterman DA, Jia Y, Martin GB.** 1996. Initiation of plant disease resistance by physical interaction of AvrPto and Pto Kinase. *Science* 274:2060-2063.

35. **Tang X, Xiao Y, Zhou JM.** 2006. Regulation of the type III secretion system in phytopathogenic bacteria. *Molecular Plant-Microbe Interactions* 19:1159-1166.

36. Underwood W, Melotto M, He SY. 2007. Role of stomata in bacterial invasion. *Cellular Microbiology* 9:1621-1629.

37. **Varvaro L, Fanigliulo R, Babelegoto NM.** 1993. Transmission electron microscopy of susceptible and resistant tomato leaves following infection with *Pseudomonas syringae* pv. *tomato*. *Journal of Phytopathology* 138:265-273.

38. **Venette JR, Lamey HA, Smith RC.** 1996. Bacterial spot and bacterial speck of tomato. North Dakota State University, Fargo, ND *Available at: http://www.ag.ndsu.edu/pubs/plantsci/hortcrop/pp736w.htm.* Accessed 03/23, 2008.

39. **Wise RP, Moscou MJ, Bogdanove AJ, Whitham SA.** 2007. Transcript profiling in host-pathogen interactions. *Annual Review of Phytopathology* 45:329-369.

40. **Wright CA, Beattie GA.** 2004. *Pseudomonas syringae* pv. *tomato* cells encounter inhibitory levels of water stress during the hypersensitive response of *Arabidopsis thaliana*. *Proceedings of the National Academy of Sciences* 101:3269-3274.

41. **Xiao Y, Lu Y, Heu S, Hutcheson SW.** 1992. Organization and environmental regulation of the *Pseudomonas syringae* pv. *syringae* 61 *hrp* cluster. *Journal of Bacteriology* 174:1734-1741.

42. **Zitter ZA.** 1985. Vegetable MD online: Bacterial diseases of tomato. Cornell University, Ithaca, NY *Available at: http://vegetablemdonline.ppath.cornell.edu/factsheets/Tomato_Bacterial.htm.* Accessed 03/23, 2008.